The God of Creation Is the God of True Scripture Text

by
Dr. L. Bednar

William Carey Bible Society

King James Bible Research Council

Accelerated Biblical Missions Institute of
Silvery Lane Baptist Church, Dearborn Hts, MI

> Emphasizing translation of the Received-Text basis of the KJB to give God's true Word to world people groups in their languages

Copyright © 2010 Lawrence Bednar

No part of this publication may be copied or electronically reproduced by any means without prior permission from the publisher.

Prologue

In two separate topics, this treatise presents evidence that God is the author of creation and the true scripture text, as expected in a world He fashioned. The first topic discusses God's creation vs. man's invention of evolution. The second topic shows that God inspires inerrant autograph originals of scripture and guides the text throughout history to apply its inerrancy to a final product, a true translation in a language of His people.

The battle for Truth

How do we separate truth from error in finding direction in life? Many follow the "experts," all of whom are fallible and often have a slanted viewpoint, being taught the bias of liberal thinkers who do all they can to spread their bias. The danger here is getting so acclimated to error that truth can no longer be recognized, which has happened in the field of science. Most well-educated people have heard evolutionism declared as science so often they think disbelief in this dogma is ignorance. Humanist scholars loudly scorn the Bible and influence many who have no real foundation on which to build their lives. Humanism is such a natural response of the natural man to life's circumstances that he easily prefers evolutionism and its denial of God and divine revelation. But it is easily shown that evolutionism is a mere figment of the imagination of those who hate God and do all they can to spread their bias. They automatically reject the spiritual dimension of life, as if nothing beyond the material world were possible. Thus they deny their own spiritual nature.

Evolutionists end up so closed to truth they reject evidence of facts in the natural world contradicting their belief, while true science considers all evidence in reaching conclusions. Indeed, science is all about laws of nature, but evolutionist philosophy doggedly and irrationally contradicts the laws. For example, the second law of thermodynamics shows that, in the absence of intelligent intervention, energy states naturally decay toward lesser order and thus lesser usefulness. Evolutionists suggest the opposite, energy naturally advancing to states of higher order and usefulness, as in the suggested biological evolution of disorderly inanimate molecules to the extreme order of human life. In terms of heat energy, this is like saying heat can flow the wrong way, from a cold body to a hot one that we might warm our soup with ice cubes or cool our fingers in molten steel. Is there a volunteer willing to test this evolutionist concept? I suggest the soup option, which is limited to ruining his appetite.

Evolutionist dogma has been perpetrated so long by so many that "Bible-college" teachers today often say that the Bible must be reconciled with evolutionism, making an invention of men's imagination their final authority. Evolutionists spread their bias as they teach legions to parrot their views as if they possessed all knowledge, when they just enjoy sneering at what they dislike.

Now the Bible isn't a science textbook, but it addresses science with flawless logic and reveals how that which passes for science has gone astray. On the other hand, evolutionists keep changing their notions of truth as they struggle to make their theories look plausible. The only way to find truth in science, or any of life's aspects, is by the guidance of a trusted, established, unchanging standard. The only standard of this type is the enduring Word of the God who created and ordered all things, and true science steadily affirms the truth of the Word. It becomes evident that the divine Creator is the originator of true science and true scripture, if one wishes to see. But many want to be their own god and reject ultimate authority, which appears to be very hazardous, for scripture tells us God has ordained a day of reckoning regarding our faithfulness to His Word.

The writer
Dr. Larry Bednar is a retired research engineer who has loved the sciences from boyhood, accepting evolution dogma early, just because science-types were supposed to believe this. When he received salvation in Christ, his perspective on science changed

entirely, along with his perspective on God and life. Many have this type of experience, for truth imparted by God's touch in salvation will always defeat error. Indeed the day comes when love of earthly matters like science no longer compares with love of God's Word. For Dr. Bednar, this took the form of a love of biblical textual studies, as appropriate, as this too is a technical subject, but one aimed at bringing the truth of God to all those desiring to know His inerrant eternal Word.

Today the work of Dr. Bednar includes presentation of seminars on subjects like recognizing evidence of God's hand on the true scripture text, recognizing biblical inerrancy, demonstrating the total accuracy of the KJB and its textual basis and contrasting the science of creation with the nonsense of evolution theory. A science background has proven useful, doubtless because of guidance by the Creator who knows from the beginning all that will be done to prepare His people for ministry that He ordains.

The text
In the creation-science topic, the language is non-technical as much as possible, but this can't be avoided entirely when trying to clarify truth in technical matters. Detailed end notes are for those having a more intense interest in science, especially from a theological perspective. The booklet may be useful to some who have heard evolution theory espoused as science so long, they've wavered in their views, and need to see how unscientific and unrealistic such theory really is, and how scientific and realistic God's Word truly is. Hopefully the booklet will prove helpful to those who profess Christ as Savior but have fallen into the trap of thinking the Bible must be reconciled with evolution theory. It is always necessary to correct men's error with God's inerrant Word, for His Word is the ultimate authority in all matters of life and eternity.

Contrasting Biblical Creation and Evolution

When did you begin to exist? You inherited your genes from your parents, who in turn inherited those of you and your siblings from your grand-parents, who in turn inherited those of you, your siblings, parents, uncles, aunts and all their offspring, from your great-grandparents, who in turn inherited those of you, your siblings, your parents, uncles, aunts and all their offspring, your grandparents, great uncles, great aunts and all their offspring, from your great-great grandparents, etc, etc. All ancestries trace back in ancient history to two original great-great-great-great-great-great-great....grandparents whose bodies contained all genetics producing all generations. And the number of generations that can derive from the billions derived from the original parents seems unlimited. Incredible genetic complexity determining hair color, muscularity, ethnicity, height, personality, intellect, etc. in all past, present and future generations was present in two original parents, so your genetics, or you (except your spirit), were created at creation of Adam and Eve over 6000 years ago. Unlimited propagation of very complex life from two original parents is due to an infinite Creator having no limitations, as the Bible teaches.

Closely-related persons could intermarry early in world history, but that changed later as mutation potential increased.

Now evolutionists suggest that lifeless slime somehow produced a living one-cell organism (a secular miracle), from which all life-forms, derived, human, plant and animal, making us distant cousins to dandelions. Supposedly, a seemingly-endless chain of incredibly complex life-forms all derived from lifeless original slime having no genetics of any kind. And incredibly complex design in higher life-forms supposedly developed from lower forms by proposed mechanisms that fail any test of technical feasibility. And proposed millions of years of species development from other species didn't leave behind any specimens exhibiting any actual change of species into other species (claims of this always prove to be bogus). Evolution defies laws of genetics science.

Laws of Science

Science is all about laws of nature. Those laws refute evolution, and they prove to be secular restatements of biblical principles, revealing the harmony of God's creation with His written Word.

Evolution and Physics

The 1st law of thermodynamics tells us total energy plus mass (mass is amount of matter) in the universe is fixed. Nothing can be added to the total, so natural processes can't create, denying the evolutionist claim that the universe and all life forms are self-created by natural processes. The law shows energy/matter exists by virtue of a power outside the universe, as in biblical creation.

The 2nd law of thermodynamics tells us energy/matter naturally becomes less ordered and thus less available and potent, or less useful, with time, so all energy/matter steadily decays, reflecting the biblical principle of death as a curse for sin. The law denies evolutionist notions on the universe as eternal or billions of years old, showing it was created rather recently, or it would be dead or approaching a state of energy rigor-mortis by now. It's very much alive and energetic, satisfying the 2nd-law criterion for youth.

The law refutes supposed molecules-to-man evolution that would involve natural increasing order, availability and potency of energy/matter. In terms of heat energy, such evolution would be like a natural permanent reverse heat flow, so that cold air might warm us in winter, or hot air might cool us in summer. Increasing order occurs only temporarily, only locally, and only by intelligent input. Nature can't do impossibly unnatural things.

Evolutionists cite natural 2nd-law reversals,[1] like order in snowflake crystals formed from disordered water-vapor molecules. But that's just one-step ordering, undone as temperature increases. Evolution needs millions of progressive steps in permanent 2nd-law reversal, and is denied by snow-crystal formation. And the Creator's intelligent water-molecule design is vital to snow-crystal temporary 2nd-law reversal. And snow-crystal formation is a downward step from a more available potent energy level that can do much work, to a less-available less-potent one that can do less work, in accord with the 2nd law. Ordering and related changes occur downward, together or separately, reversing only temporarily, only by intelligent intervention and only by one step in snow-crystal formation, as required by the 2nd law.

Evolutionists cite 2nd-law reversal by the ram pump that uses a valve to elevate water flowing downstream, increasing potential energy availability and potency. But this is a non-progressive reversal, and there's a net loss in energy availability, energy gain being less than that expended to elevate the water, in accord with the 2nd law. And intelligence is needed to design the pump, and the pump wears out, so new intelligent input is needed to continue resisting the law. No permanent progressive 2nd-law reversal by unintelligent natural means is known and can't be justified.

Both laws of thermodynamics are secular restatements of biblical principles, taking us nearer to God who created the universe and endowed the Bible. The 2nd law is His mark on a fallen creation, death as the curse for sin encompassing man, animals and the universe at large. The 1st law speaks of God's creative newness and what might have been, and someday will be, eternal life in a world that never decays, and the 2nd law eternally repealed.

Evolution and Biology
Laws of genetics: Evolutionists defy life-only-from-life biogenesis law, saying inanimate molecules united at random in slime to yield a tiny living cell. From this came all plant and animal species, making us distant cousins to slime, bacteria and weeds. They postulate such things, for rejection of the supernatural gives them no other way to explain the origin of all the forms of life.

They defy law limiting species reproduction to each species, saying from the initial cell came intermediates combining aspects of certain species, and from these came each species. Thus creatures with no bones, fins, wings, lung-respiration, complex brains, etc. supposedly sprouted these bit-by-bit over the ages. Proposed causes of this magic are mutations (which are destructive, not constructive) or isolation of groups from others of their species.

Now if laymen observe reproduction gone berserk, non-living slime producing a living cell that produces living creatures, this is called a horror movie. If evolutionists imagine reproduction gone berserk, slime producing a living cell that produces a series of creatures from bacteria, to fish, to frogs to snakes, to dogs, to apes, to people in gradual change over some millions of years, all in defiance of science laws, this is called science.

In the Bible, created life continues to derive only from life, and each kind produces only its kind, in accord with laws of genetics. Evolution isn't even good theory, for a theory is a <u>plausible</u> unproven explanation for reality, but faith in evolution has no basis even in theoretical science and is really a secular religion. Yet evolutionists accuse Bible-believers of religious mythology.

Genesis
1:12 *And the earth brought forth grass, and herb yielding seed after his kind, and the tree yielding fruit, whose <u>seed was in itself, after his kind</u>...*
1:24 *And God said, Let the earth bring forth the <u>living creature after his kind</u>.*

Genesis reflects laws of genetics, and both deny evolution theory on kinds producing other kinds. *After his kind* limits propagation of a kind to members of the kind (we clarify kinds vs. species later), in accord with laws of genetics. Evolutionists defy this, saying kinds develop from other kinds by radical genetics unknown to true science. They've said modest species variance like different dog types indicates radical long-term evolution, defying laws of genetics that show species variance restricted to modest limits defined by a unique gene pool inherent to each species.

Early-evolutionism's "simple" human cell: A 19th-century pre-scientific notion that drove evolution theory is the suggestion the human cell is just a blob of protoplasm open to evolution. But 20th-century study revealed mind-boggling cell complexity and orderly processes that make the idea of biological evolution a fantasy. The cell is part of God's creation, as seen by Psalm139.

Psalm 139
By God's Spirit, David tells of the complexity of man.
14. *...I am fearfully and wonderfully made: marvelous are thy works...*
15. *My substance was not hid from thee when I was made in secret, and curiously wrought in the lowest parts of the earth.*

David tells of amazing human-life complexity arising from the lowest parts of the earth, creation from lowly dust, and life propagation by mankind's lowly reproductive systems, to reveal the limitless creation power of God. He brings incredible complexity of life from even the simplest and least noble parts of the earth.

The "Ape-man:" Where is he today? Scopes Monkey Trial: This 1925 court trial began a move toward evolution as the one theory of origins allowed by public-school boards. A technically-unqualified W. Jennings Bryan was embarrassed by recent discoveries like "Nebraska Man" representing supposed evolutionary development. Nebraska Man proved to be a case of ignorance fabricated from one old tooth, said by an "expert evolutionist" at the Scopes trial to be that of an intermediate to man and apes. Later the tooth was found to belong to an ancient pig.[2] We can sum-up the matter by saying, evolutionism pigged out on expert ignorance and stupidity, plus mass school-board gullibility.

"Ape-men" specimens always prove to be bogus.[3,4] Some noted below show how incredibly unsound the ape-man teaching is.

1. Piltdown "Man" This ape-man, used at the Scopes trial, was in science textbooks for over 40 years before it was found to be

bogus. It proved to be a human-skull portion and an ape's jaw and teeth, stained to make them look old. The teeth had been filed to make them more human-like, and tools and fossils were imported and planted. In other words, evolution's stained jaws and teeth filed a claim in the stained human skull, and the ape-man was imported from imagination and planted in books.

The Piltdown mentality lives on in a pro-evolution national-magazine report on a Chinese specimen with supposed morphology of birds and dinosaurs.[5] The specimen was bogus, crudely joining bird and reptile fossil parts. Noted evolutionists advised against publishing the report, but blind faith evolved a deaf ear.

2. Dr. Eugene DuBois' Java "man"
Dubois set out by faith to find an intermediate. He found an ape-like skull cap, associating it with a human thigh bone found later ~40 ft. away. He hid two human skull caps from his findings, and others discovered modern flora/fauna fossils in the soils that DuBois studied. It looks like this eager advocate of evolution couldn't connect the thigh bone to the head bone.

3. Louis Leakey's "Ape man"
African ape-like bones and skull were said by Leakey to belong to evolving man, due to the finding of tools nearby. He made a case for an upright walking gait and teeth of some human character (despite strong ape-like character). His results, praised at first, are now disregarded, rebuking his blind faith. His son Richard found human bones under those he found, explaining the tools.[1] Being under the "ape-man" remains, and so a little lower in the geologic column of slowly deposited soil strata so popular in evolutionism, the human bones would be older than those of man's "ape-like ancestor." Thus blind "science" aped blind faith.

4. "Peking man"
At a site near Peking, ape-like skull fragments were joined to face bones from a few feet away and a human-like jaw from ~80 feet away, yielding a skull looking like that of an intermediate. Later the site proved to be that of an ancient hunting party, a few

of which died and left human bones behind. They killed monkeys, throwing the skulls and bones in a fire. Skull fabrication involved piecing together remains of skull bones of monkeys and other animals and human-skull remains from a more remote location. Thus evolutionism fabricated bones of contention.

Now if laymen promote disinformation, this is called a hoax, but if evolutionists do so, this is called adjusting facts.

In contrast with blind faith, Christian faith is based on visible evidence from the invisible realm (Heb.11:1), results of the new spiritual birth and God's guidance in vital matters of earthly life.

Wanted: A mechanism to power evolution
Only supernatural creation can account for a sudden full-blown appearance of all life, so evolutionists need ape-men to suggest gradual transition among species as basic to evolution theory. The greatest problem has always been identifying a mechanism by which natural processes could do such unnatural things.

The imagination of blind faith: The search for an evolution mechanism has been a comical pilgrimage. An early suggestion was slow species change by anatomical exercise, such as a long-necked giraffe-like species evolving from a short-necked zebra-like one as the latter reached higher for edible tree leaves as food supplies dwindled in a drought. This is like saying a weight-lifter genetically transmits a tendency toward large muscles, which laws of genetics deny. A child won't be genetically inclined toward larger muscles any more than he'll be genetically inclined toward broken arms if dad breaks his arm. Muscles develop if a child follows his dad's interest in athletics or hard-work habits. Environmental effects on genetics are just degenerative ones of mutation and minor ones of adaptation within gene-pool limits.

A dated basic mechanism, still offered in desperation, is gradual transition to new species by cumulative mutation. It's said nature favors transitions yielding viable organisms, other types dying out. But mutations are destructive, and if constructive ones exist, they're extremely rare. How can something so unlikely to occur once, occur millions of times in succession without interruption to yield one real organ of one human or animal body?

The notion of evolution by mutation was due to observations like insect adaptation to DDT and flu-virus adaptation to vaccine. But this minor variance is a species' gene-pool provision against extinction, not an evolutionary step. No evolution development occurs at the limits of variance allowed by a gene pool, as seen in interbreeding of male donkeys and female horses to yield mules. The closely-related donkey & horse are mildly-different species, (according to our less-than-perfect definition of species), and the normally-infertile mule is the variance limit. Even a mild donkey-horse difference won't develop beyond the mule, and less-related creatures can't interact reproductively at all. Furthermore, mutations only deteriorate a gene pool. Gene pools don't allow the basic anatomy/physiology change of evolutionary transition.

In the search for a transition mechanism, much attention is paid to interbreeding irregularity of some members of a species. Interbreeding ability is basic to species identity, and evolutionists say a lost ability of this type identifies a new species. They note that environmental isolation of fruit flies (and birds or lizards) results in the loss of interbreeding ability with outsiders of their own species. But this loss in environmentally-isolated flies is at most a variance within gene-pool limits as an isolated group adapts to restricted mating possibilities. It may even be just mating preference due to variant morphology from restricted interbreeding, and so may have no effect at all on reproductive ability.

Now if men lose reproductive ability, this is called sterilization, but if flies lose that ability, this is called evolution of species.

As pointed out,[2] loss of interbreeding ability is a step downward to extinction, not upward in evolution development. The survival potential of an isolated group of a species is reduced by fewer members for propagation. Isolation reduces species gene-pool variety, defeating evolution theory. Evolution demands expanding gene pools with new capabilities to account for "increasing organism complexity with species upward development." To argue for species generation by isolation is to offer increased extinction potential as the driving force of evolution, and this is just an effort to ensure survival of out-dated evolutionism.

Evolutionist faith fails: But suppose a religious-type miracle of natural processes somehow caused transition. If all species derived from a common ancestor, we would expect, as Darwin did, to find billions of intermediates exhibiting every tiny degree of anatomical mixture of certain different species as transition progressed very slowly over millions of years. For fish and frog types, we should see a superabundance of "fishog" and "frogish" types, of every degree of mixture of the two, exhibiting very slow change among fossils and living creatures (has anyone ever seen any suggestion of a fish tail on a frog, a fish that croaks, or a fish with frog legs?) Intermediates would outnumber developed types by huge proportions among fossils and living creatures, but not one real transition has ever been found in the fossil record (despite evolutionist claims) or among living creatures. If evolutionary transition is a fact of science, where are intermediates that should be superabundant in the fossil and history records? Why in the world is there no evidence of their existence?

Evolutionists point to mosaics as supposed transitions. But a mosaic just combines features usually confined to different species, no feature showing evidence of evolving into another (such as a reptilian scaly limb developing into the feathered wing of a bird). Evolutionists have never found a single body feature of one species showing a partly developed change into a body feature of another species, denying evolutionary transition.

Even more telling is lack of transition evidence at the DNA/protein molecular level of life. Evolutionists say different species can be linked in a development sequence by morphology ties, and development supposedly progressed from fish to amphibians to reptiles to mammals. But the latter three are seen to be equally isolated from fish at the molecular level, so evolutionary development isn't possible.[6] There are no molecular-level ties to prove evolution from one form to another, and each class of life is unique and isolated from others at the molecular level. There is even a near-perfect mathematical order to the observed degree of molecular-level isolation among the divisions of life, as expected of creation of distinct kinds. This isn't expected of evolutionary transition that should induce many combinations in new complexity. A lack of molecular-level ties between species said to be morphologically related, fully discredits evolutionary transition.

Molecular-level studies dispel the notion of mosaics as transitions. An example, the lungfish, has gills for underwater life and lungs for life in periodically-dry ponds of arid regions. The lungfish shows no molecular-level evidence of transition,[7] so its dual capability is a creation provision for survival in its habitat.

Life's pattern is distinct kinds, and the case for transition is hopeless. All frogs are morphologically close, but are as diverse at the molecular level as mammals, that are so morphologically diverse as to include people, whales, bats, moles and cows.[7] How can evolution build new species from others when molecular diversity producing only various frogs must produce people, whales, bats, moles and cows? How can morphology, so dependent on the molecular basis, begin to advance with no advance in the molecular basis, and how can random evolution processes make this happen? Evolution demands transition, and there are no molecular building blocks crucial to transition. Ingenious God-ordained creation alone accounts for the extreme morphological differences in species of comparable molecular diversity.

Now if laymen propose notions contradicting facts, this is called stupidity. If evolutionists do so, this is called faith in science.

Evolutionist faith keeps failing: Some still say a mosaic like the lungfish, or a morphological similarity like that of whales and sharks, reveals transition,[8] but, with no evidence of gradual transition, that means existent species would be transitions. Do they think fully-developed species have arisen from other fully-developed species, which would require interbreeding of a wide variety of species, in contradiction of the meaning of *species*? In their desperation, evolutionists have done so, as we'll now see.

Some evolutionists now forsake gradualism, favoring the opposite, abrupt evolution. A discarded early theory, Hopeful Monster (H.M.), suggested things like two reptiles hatching something like a bird.[9] A related new theory, Punctuated Equilibrium (P.E.), suggests rapid short intermittent evolution spurts [9] in isolated groups with poor survival potential, short life and few members accounting for lack of transition evidence. Geographical/social factors in isolated groups prevent interbreeding with outsiders of their species, producing a new species by "splitting of lineage." But again, environmental isolation allows only degenerative mutation or minor variance within gene-pool limits, and reproductive isolation makes greater extinction potential and a lesser gene pool of a small group, evolution's driving force.

P.E. is just a new spin on discarded H.M. Both eliminate the need for transition and yield a radically foreign creature that miraculously finds a like creature of opposite sex to propagate itself. P.E. depends on impotent environmentally-based genetic innovation, and H.M. on impossible genetic innovation. P.E. defies science law by impotence, and H.M. by impossible monstrosity.

With evolution defying science laws, advocates now say we can't assume the laws always applied as they do now in our part of the universe, which is imagination. We can't postulate beyond what we observe in nature or the Bible through God's revelation.

Now if laymen relate dreams to reality, this is called fantasy, but if evolutionists do so, this is called expanding science frontiers.

Hopeless evolutionist faith: Evolutionist biology fails all tests. The bombardier beetle has, in body sacs, flammable liquid and oxidizer that react if mixed, and are ejected in a fiery emission for self-defense. If just one chemical evolves, self-defense fails, and the insect is soon extinct. But suppose he's incredibly lucky, beating all the odds so that he and both chemicals all evolve at the same time. He will soon be extinct due to impossible odds against another beetle just happening to evolve itself at the same time and happening to evolve both chemicals at this time and just happening to be of opposite sex to permit propagation of the new species. But if the beetles manage this, they still become extinct by immolation as the chemical mix ignites in their bodies, with no inhibitor normally controlling reaction having evolved at the right time. Evolutionary development would ensure extinction of these incredibly incredibly lucky, but very unfortunate beetles.

And can an incredibly incredibly incredible evolution of the insect and all three chemicals just happen at the same time, and at just the time another beetle just happens to evolve itself and all three chemicals together at the same time and just happens to be of opposite sex for the sake of propagation and self-defense? This is so improbable it would be a statistical impossibility over an essentially infinite time. And with self-defense and reproduction being just part of the insect's anatomy/physiology that must function as a whole for survival, creation with all body parts and systems fully functional, is the only possible explanation for the insect's existence and survival.

Now if laymen expect magic in nature, like a pile of gold at the end of a rainbow, this is called a fairy tale. If evolutionists expect magic in nature, like a pile of species at the end of a molecules-to-man chain of fairy-tale genetics, this is called science insight.

Evolution and the Age of the Earth

Evolutionist faith undeterred by reality: Millions or billions of years stipulated for evolution suggest its plausibility, but time denies major tenets of evolution. The 2nd law of thermodynamics says all things simplify with time, contrary to natural increasing complexity needed for major aspects of evolution to be feasible.

Evolutionist age estimates of millions and billions of years involve potential error that is never mentioned. Decay of radioactive isotopes in igneous rock formed by lava solidification, produces a stable element, all of which found in rock is attributed to isotope decay. Other sources can contribute, and tiny amounts will inflate estimates greatly since that produced in millions of years by isotope decay at current rates would be very small. And age is calculated assuming a constant historical decay rate, which is unproven (evidence indicates rates were once much higher).

Evolutionists ignore age data that doesn't suit their theories. When accuracy of isotope-decay methods can be checked, estimates can prove to be vastly inflated. Morris [10] reports results at Mount Kaupelehu in Hawaii on rock formed from 1800-1801 volcanic eruptions. Twelve estimates by various isotope methods ranged from 140 million to 2.96 billion years, or ~830,000 - 18,000,000 times known age. And the 21:1 variability factor for different methods even reveals a variable degree of inaccuracy.

Morris reports on the popular decay of potassium-isotope to argon-gas method. Estimates on rock formed on the ocean floor by lava flow at Mt. Kilauea in Hawaii ~200 years earlier, were as great as 21 million years, ~100,000 times actual age. All rock was from one lava flow, yet age varied with rock depth in the ocean, indicating water pressure hindered argon escape in rock in deep water, inflating estimates. Rock from the least depth had argon content too low to measure, indicating a more realistic age.

Igneous rock at Mt. Rangitoto, New Zealand, dated less than 300 years age by radiocarbon studies on trees destroyed by eruptions forming the rock, had a 485,000-year potassium/argon age (~1600:1 variance for different methods). Rock age at Salt Lake Crater in Oahu Hawaii, was less than 400,000 years with one method, but 2.6 million - 3.3 billion years with others (variance for different methods as much as 8,250:1). In rock from a Grand

Canyon volcano, the potassium/argon age was 10,000 and 117 million years for different minerals in rock from one lava flow known to be of recent origin (extreme inflation plus 11,000:1 variance in one method for different parts of one rock specimen). Different minerals retain different argon amounts, further complicating and distorting age estimates by such a method.

When estimates prove as inaccurate as they were in these cases, all isotope-based measurements become highly suspect.

Now if laymen ignore inconvenient facts, this is called dishonesty, but if evolutionists do so, this is called scientific selectivity.

No compatibility of Christianity and Evolutionism

Scripture denies evolution: Evolutionism is secular religion, a faith in nature, and a hedonism denying sin's penalty. Many try to reconcile biblical creation to evolutionism (Scofield-Bible notes offer Gap theory). Evolution is unbiblical unscientific dogma, *science falsely so called* (1 Tim.6:20). God created the universe and endowed the Bible, so true science will agree with scripture.

Gap theory: Some say a first earth perished in a Genesis 1:1-1:2 time gap, causing a Genesis 1:2 formless, void dark earth, so the 6 days following would be a re-creation. They say Genesis 1:2 reads, *And the earth became*, not *was, without form. Was* is right translation of the Hebrew, for an earth ruined (by satan's sin) doesn't fit in God's view of the creation as *very good* (Gen.1:31).

Gap theory is far-fetched, supposing missing words between verses 1 and 2 (or before verse 1) to create a knowledge gap. Why would a matter so basic to understanding the creation account, and so far beyond deduction textually, be omitted?

The theory addressed supposed millions of years of rock-strata age surmised by circular reasoning.[11] Fossils are dated by ages of strata containing them, and strata are dated by fossil ages assumed in evolution theory. An index species is chosen on a supposition it became extinct millions of years ago, its fossil in strata marking the strata as millions of years old. But an index specimen was recently found alive, so its fossil can't prove strata age much greater than the life span of one index specimen.

The Gap theory conflicts with scripture. As has been pointed out,[12] fossils in the geologic column would suggest life and death millions of years before Adam. The Bible says death entered the world by Adam's sin, and he didn't live until the 6th creation day. The theory alleges life and death in a period of millions of years between Genesis 1:1 and 1:2, and so before Adam and before the 1st creation day. To improve theory plausibility, it's said human-like creatures before Adam had no soul, but they would be like animals, and animals didn't die before Adam's sin (Rom.5:12 – plant life didn't die in this sense, having no conscious existence).

Exodus 20:11 says God made all things in 6 days of Genesis 1:3-26.[12] But Gap theory suggests an earth destroyed and producing fossils before Genesis 1:2, long before the 6 days began.

Other ideas: Theistic evolution suggests God ordained evolution in a self-progressive or guided-creation process, and the day-age notion suggests God's guidance and periodic intervention in creation days that each signify an eon. Thus millions of years of death before Adam are postulated, long before the Bible says it entered the world by sin in Genesis 3. And the Bible says God created in a week of days, not eons, as we'll see. And evolutionists say earth arose after the stars, but scripture says earth was formed first. We can't be both Bible-believers and evolutionists.

The Logic of Creationism and a Young Earth

Leaving the joke of evolution, we consider the reality of God's creation. True science relates to the Bible, for both originate in God and can't disagree. Men didn't observe the creation, but the Bible informs us about it, and true science is supportive.

The Bible shows Creation days were 24-hour periods, not eons. Exodus 20:9-11 says, *Six days shalt thou labor, and do all thy work: But the seventh day is the Sabbath of the Lord thy God: in it thou shalt not do any work...For in six days the Lord made heaven and earth, the sea, and all that in them is, and rested the seventh day: Days* are 24-hour periods in a 7-day week. And *morning/evening* in each creation day indicates 24-hour periods. And Gen.1:14 relates *day* to 24-hour periods, saying heavenly bodies are *for seasons, days and years*. And Gen.1:16 calls the sun a *light to rule the day*, the 24-hour period. It's said there was no sun to determine days until Gen.1:14, after creation days began in 1:5, but Christ's divine glory was the first sun (2 Cor.4:6).

Creationism's heaven and earth: Scripture says God began the creation in a preliminary, not ruined, state. This was water vapor (Gen.1:2), accounting for a lack of form and features. A huge formless water-vapor cloud fits this description and accounts for waters above and below a dividing firmament (Gen.1:7).

Water vapor under the firmament was *gathered together*, or compressed for condensation to form liquid seas. Water above the firmament is vapor in a concentric canopy above the seas, *above* being, away from the center of gravity, and condensation of this canopy would be part of 40-day rain of the great flood. Liquid water under the firmament would naturally be a sphere due to gravity effects, but solid earth growing within the sphere would displace water to the surface to create oceans in surface valleys. Earth's hot core and overlying molten layer would form through heat of temporal accelerated isotope-type decay in the extreme energy conditions of the first two creation days.

A water-vapor canopy is indicated by *windows of heaven* opened in flood rain (Gen.7:11), signifying canopy collapse as a cause of earth's first rain (Gen.2:5). Canopy loss explains a post-flood loss of great human longevity, the canopy shielding against excessive harmful solar radiation (UV, gamma, X-rays) and its longevity-limiting mutations and cancer. And a canopy above the 1st firmament explains a uniformly-warm early earth indicated in fossil studies,[15] a greenhouse effect ending with the flood. Extreme earth heating expected of a thick canopy could be negated by a heat-shield effect of ice crystals in the upper canopy.[16] Most likely, canopy thickness was modest, and evaporation of warm groundwater in erupting *fountains of the great deep* (Gen.7:11) caused by tectonic activity, contributed much of the flood rain.

Limited accelerated radioactivity accounts for energy release in tectonic/volcanic activity implied by great fountains of the deep breaking up in the great flood (Gen.7:11). Ocean warming and much evaporation adding to flood rain would result. And heat of accelerated decay could contribute to the energy release that will underlie the fire of transformation that is to change the earth and the sky at the end of the Millennium (2 Pet.3:7-10).

The Gen.1:8 *firmament* is *heaven*, the atmosphere with a water vapor canopy above and water vapor to be liquefied below. Gen. 1:15,17 *firmament of the heaven* as the sun and moon locale can seem to place the sun and moon in the atmosphere. But Hebrew *heaven* is a dual-no. noun that signifies two heavens, each with a firmament. The 1:8 *firmament* is *heaven*, the atmosphere, and a 1:15,17 *firmament of the heaven* is a 2nd firmament of the 2nd heaven, outer space (a Hebrew construct-possessive link signifies a firmament that belongs to heaven). Firmament is substance, and air (gas) filled 1st heaven, so this firmament is 1st heaven, but 2nd heaven is largely empty space plus a firmament of this heaven (gas, dust, stars). Other scripture notes two senses of *heaven* (Dt. 10:14, *the heaven and the heaven of the heavens*), a distinction lost by the modern-version *expanse*. *Expanse* replaces *firmament* in 1:8, and *expanse of the heavens/sky* replaces *firmament of the heaven* in 1:15,17, suggesting one heaven located between the waters, and the sun and moon located in the atmosphere (the construct-objective link is incorrect since *expanse* indicates the extent of just one heaven, ignoring the noun dual no.)[13,14]

1st-heaven *firmament* substance is air to enable respiration, support aircraft/birds, create wind, control earth temperature, etc. Beyond is a 2nd-heaven firmament with thin water vapor, a substance likely involved in creation of heavenly bodies.[17] Water and its constituent hydrogen are widely dispersed in outer space in the sun, other stars, meteors, comets, planets, gas clouds, etc.[18,19]

Third heaven, God's dwelling place, is the basic *firmament*. Like the atmosphere and outer space, the invisible third heaven seems to lack substance, but it's the true substance of everything, the firmament that is the locale of creation power.

Creation Biology: Laws of genetics show creature reproduction outside species limits is nullified. Scripture says God made each type of animal/plant life after its kind, limiting propagation of each to its kind. Groups of related species can be a *kind*, as in Lev.11:13-19 that includes in a *kind* various bird relatives like owls, ravens, hawks and vultures. While these birds are related, they can't interbreed, so when scripture say propagation here is limited to a kind, it's saying birds produce only birds.

But scripture also makes the term specific, extending *kinds* to subtle distinctions of the modern concept of species. Lev.11: 22 denotes as different kinds, closely-related grasshopper, locust and bald-locust species, so a biblical *kind* can distinguish true species according to today's definition. Biblical biology is quite accurate, unlike evolution's magical intermediates or H.M./P.E.

Complexity in the DNA/protein heredity basis reveals the reality of inhabitants of Noah's ark as the source of species diversity seen today. A near-infinite variety of minor human features and traits like nose size, hair color and disease inheritance are found in the gene pool of one set of parents. A reputable geneticist/ evolutionist says the gene-pool variety of one couple can account for all variant traits of humanity,[2] being theoretically capable of producing as many as $10^{2,017}$ offspring before two identical ones result. This many people are many more than can be contained in a universe trillions of times larger than that known to us (10^{12} would be a trillion offspring). Animals too have an extremely large gene pool and much minor variety in one set of parents.

Yet evolutionists reject origination of today's human and animal diversity in the post-flood four couples and many animal pairs.* They say unlawful radical transition spawns new species, yet scoff at ark lawful genetic diversity to account for variety in species today, and this after using genetic diversity to justify evolution transition (pure hypocrisy). They artificially increase the number of species by inventing differences within a species to suggest there were too many to fit at least a pair of each on the ark, but interbreeding ability was the only determining factor.

Genesis 8:15,17
15: *And God spake unto Noah, saying,*
17: *Bring forth with thee every living thing that is with thee, of all flesh, both of foul, and of cattle, and of every creeping thing...that they may breed abundantly in the earth, and be fruitful, and multiply upon the earth.*

Creationism's young earth: Earth age indicated in scripture is ~6000 years, based on genealogies. This is a rough estimate, and a likely maximum is 8,000-10,000 years. Huge potential error in isotope-decay estimates mandates use of more realistic methods.

Some tree fossils in coal mines extend in their length through coal and sedimentary-rock layers that each supposedly required millions of years to form from organic and mud deposits.[20] But up-rooted dead wood, rotting away by exposure to water or air, can't maintain its form for millions of years as each layer forms around it from slowly hardening deposits. Clearly, layers formed over a period of time short enough to allow a dead tree portion to maintain its shape throughout the period, so the period was brief. This indicates the strata formed quickly from rapidly deposited organic and mud sediments in the great flood ~4400 years ago, so that's how old the rock and coal are (Water as ice now covers

---

*Intermarriage exclusion of close family members is due to mutation potential, not yet a problem at Noah's time, explaining how, ~450 years later, Abraham could marry a half-sister (and how Cain could have a wife).

high Antarctic mountains,[21] and marine fossils are seen on mountains all over the world, so water once covered them, as scripture says - they would be much lower before the flood, being elevated later by protracted volcanic/tectonic activity – Ps.104:6-9).

Brittle rock strata formed from hardened mud deposits are at times bent sharply without cracking expected if bending occurred in the hardened state.[20] Evolutionists say layers deposited and hardened over millions of years, but that would mean brittle rock layers bent sharply without cracking, which happens only under unique conditions and in minor degree. The lack of cracking happens while strata are pliable, so they formed in a short time, not millions of years. This indicates short-term mud deposition expected under the Great-Flood conditions of ~4400 years ago according to the Bible, and that's how old the rock strata are.

Regarding animal fossils, soft flexible tissue and blood cells were seen in fossilized dinosaur bone estimated to be ~68 million years old.[22] Such tissue decomposes too rapidly in fossilization to endure many thousands, let alone millions, of years exposure. Preservation due to rapid burial in sediment associated with a large flood, will be limited to a few thousand years, relating the find to the time of the Great Flood or a more recent local one. Evolutionist efforts to disprove soft tissue have been refuted.[22]

Regarding earth's mineral deposits, lead diffusion in zircon increases with increasing temperature. Temperature increases with increasing earth depth, so deeper specimens will eventually show greater lead diffusion. There's no measureable difference in lead diffusion between zircon crystals taken near earth's surface and those at depths up to 4000 meters.[23] Earth is too young to detect a difference, and thus far younger than evolutionists claim.

Decay of C^{14} isotope at the current rate is too fast to be used for age estimates over ~80 thousand years. Evolutionists say coal deposits are many millions of years old, but if that were true, no detectable C^{14} should be present. Creationists found significant C^{14} in coal indicative of age on the order of 4200 years, in accord with coal formation in Great-Flood chronology of the Bible.[24] C^{14} levels were fairly uniform at all depths in the earth's crust, consistent with coal formation from vegetation deposited in the short Great-Flood period. Evolutionists suggest C^{14} contamination of porous coal, but creationists found significant C^{14} in diamond, a form of carbon impenetrable to contamination.

Regarding cosmology, meteorites fall on earth steadily. Now if earth's sedimentary layers, that average about a mile in thickness over the continents, were deposited a layer at a time over many millions of years postulated by evolutionists, much iron/nickel-rich meteorite debris would be found in layers well below the surface. But it's found only in soil layers near earth's surface, so sediments were deposited recently, and earth isn't very old.[23]

Comets will be about as old as the solar system, and many have short-term orbital periods of ~200 years. They lose dust and ice with each orbit around the sun, producing their bright coma and tails. Disintegration rates of short-term comets indicate their lives as true comets are ~10,000 years at most.[23] A solar-system age greater than this will leave it depleted of these objects. There is no shortage, so they don't seem very old, indicating the solar system isn't very old. Evolutionists deny this, suggesting a group of ice/rock bodies beyond Neptune is a storehouse renewing the supply, their imaginative type of excuse for inconvenient facts.

Astronomers claim a universe several billions of years old, due partly to changes in star color and size in periods supposedly of ~1 billion years. Certain observations contradict this.[25] The star Sirius B shows evidence of a "billion-year" change in ~1000 years, suggesting universe age is nowhere near as great as postulated. The star Betelgeuse has shown color changes in recorded history, a matter of centuries, not a billion years. Such "mystery" baffles astronomers, but they never reject evolution theory.

Creationism's apparent earth age: An evolutionist assumption that all lead in rock results from extremely slow uranium-isotope (U-238) decay, can suggest billions of years of age. But even if all lead derived from U-238, that can mean decay was so rapid in creation that an earth with an appearance of old age resulted. God created Adam and Eve as mature adults, and earth should be created that way, geologically mature with mild tectonic/volcanic activity suited to support of life. Earth would be created with geological maturity by initial energy conditions so extreme that processes later requiring millions of years, occurred in days.

Accelerated earlier decay of radioactive isotopes would give misleading apparent great earth ages, accounting for inflated age estimates by radioisotope methods. It could produce great heat, accounting for the molten layer around earth's core. But heat/radioactivity of decay accelerated enough to account for earth age on the order of 6,000-10,000 years is expected to be so great its effects had to be limited during very energetic creation phases and had to subside rapidly for life to exist in the initial earth.

How the heat/radioactivity of such accelerated decay could subside is unclear, but studies reveal early accelerated decay.[26] Helium as alpha particles (helium-atom nuclei) in isotope decay is quickly lost by diffusion in zircon crystals since helium atoms are very small and non-reactive. But zircon shows high helium concentrations, indicating very rapid early isotope decay and inadequate rock age for loss by diffusion. A helium clock records rate acceleration, while a uranium clock can't. Studies indicating extreme early acceleration of decay reveal rock age, supposedly ~1.5 billion years, as within a 4000-14,000 year age range.

Accelerated decay is further indicated by uranium/polonium radiohalos,[27,28,29] spherical damage points in the microstructure of mineral crystals caused by ejected decay particles. U-238 decay forms intermediate isotopes that decay to form polonium isotopes. U-238 decays so slowly as to require ~100 million years at today's decay rate to give visible halos. Polonium decay quickly produces halos, and these are at times a very short distance from the uranium source of polonium and intermediate isotopes. Intermediate isotope atoms and their polonium-product atoms migrate from parent uranium by hot-fluid transport shown to exist briefly in granite rock formation. Formation of halos from fast polonium decay, and brief fluid-transport times of intermediates and their polonium product, show uranium decay had to be much faster than it is now. At the current rate, very few intermediate atoms could form and migrate in the brief fluid-movement time, and polonium halos associated with fluid transport could not form. Rock with polonium and uranium would form by solidifying of magma generated by heat of accelerated radioactivity melting sedimentary rock originating from Great-Flood mud deposits.

Early accelerated decay explains age-estimate disharmony among isotope methods, different degrees of initial acceleration distorting estimates differently. Evolutionists insist on decay-rate constancy, but even chemical/physical factors affect decay rates a little, so the notion of constant rates isn't good technology.

A question: Great universe age is suggested by distant galaxies. Light speed is ~186,000 miles/second, but some galaxies <u>seem</u> so distant their light seems to require billions of years to reach us (indirect methods based on assumption are the only ones for very distant bodies). Their visibility implies their light has traveled toward us for eons, suggesting they have existed for eons.

Creationists respond with the apparent-age concept. But at its current speed, light from a star in a distant galaxy could travel just a tiny fraction of the distance in the 10,000 years maximum scriptural earth age. It's said the concept requires creating a light beam in close proximity to earth, which is incredible, refuting the concept. And light of a star contains technical information originating at its surface, denying creation of light beams in space.

Another complication arises in the case of exploding nova/ supernova stars. In this case a light beam created in space close to earth would include an imprinted illusion of an exploding star, and there would be no actual explosion. It's said distant stars of this type exploded billions of years ago and existed that long ago.

Response: All this distorts the apparent-age concept. Gen.1:17 says stars were created to give light on earth, so their light reached earth quickly. At extreme speed, delay for very distant bodies would be a few centuries or millennia, and in a young created universe, light of extreme speed originates at its source, that of nearer stars quickly reaches earth, and novas explode. To assume light speed is constant is illogical, for it's known to decrease by interaction with matter, and space contains much gas and dust.

Apparent great age due to early extreme light speed is rejected, but studies support it. In 1927/1931/1934 reports, astronomer M.E.J. Gheury de Bray [30] noted consistent decreases suggesting earlier acceleration in 22 measurements in a 75-year period in the latter 19th and early 20th centuries. Techniques weren't refined then, but values decreased continually. Experimental error appears as variance above and below a real value, not steadily lower. A trend of decreasing speed is seen in measurements in a 66-year 1874-1940 period using reasonable techniques.[31] Physicist Barry Setterfield plotted historical light-speed data and found consistent decrease in the last 300 years,[32] but results are discounted due to uncertain statistical analysis when technique accuracy varies.

Studies indicating decreasing light speed are dismissed too readily. As Gheury de Bray said, progressive decrease can't be dismissed on the basis of experimental error. Even the last 300 years is a minor part of 6,000-10,000 years of earth age indicated in the Bible, and attenuation can be advanced enough that recent change has been so minor that values seem constant (exponential attenuation of energy itself is normal: steadily diminishing magnitude decrease eventually approaches zero so that the magnitude approaches constancy). Near-constant speed over the last 300 years is expected if the decrease occurs in this common fashion.

If results are grouped by technique, consistent evidence of a speed decrease is seen.[33] To illustrate this further, despite variant experimental error in using different equipment, 5 measurements made in a 52-year period by the rotating-mirror technique[34] show consistent decreases (299,910 km/sec in 1880, 299,860/299,853 in 1883, 299,796 in 1926 and 299,774 in 1932). Decreases aren't fully proportionate with time intervals, as expected with early methods, and there's too much consistency to ignore the trend.

If light speed decreases steadily, older data can be revealing. A treatment of 1675 A.D. data based on the apparent revolution period of a moon of Jupiter[35] showed no evidence of light-speed, decrease, but a method subject to variables of orbital motion can be too insensitive to tiny changes to be useful, and results contradict 1878-1880 Harvard data likely to be reasonably accurate.

Evolutionists say current light speed is ultimate to sustain their view. But an astronomer finds quasars expand at speeds up to 10 times that of light,[36] indicating current speed is not the ultimate.

Light speed measurements of the 1960's showed a halt in light-speed decrease,[37,38] and a real halt indicates a real decrease. The halt reflected 1960's use of atomic clocks to measure light speed. These reflect a vibrational period in an atom's change in energy state, but earlier dynamic clocks reflected earth's revolution period around the sun. Vibrational-frequency change paralleling, and exactly correlated with, light-speed change explains the halt in light-speed decrease. This is not unexpected, for a changing energy state of the atom is what is said to produce light.

This further evidence of decreasing light speed shows the concept can't be dismissed. Indeed it's even further supported by evidence of past accelerated isotope decay, these "constants" (and others) being related through properties of the atom.

Historically-constant light speed and isotope-decay rates are, assumed from inconclusive tests. Proof of acceleration of early isotope decay indicates early light-speed acceleration. Some evolutionist physicists proposed early accelerated light speed in late 19th - late 20th century reports.[39,40,41,42,43,44,45] Atom properties in creation likely were unique, and constants can't be assumed.

It's said older light of distant bodies would show evidence of any extreme early light speed, lack of this showing light speed hasn't varied. But creation with earth and man at its focus should invoke constancy in natural phenomena from an earthly perspective.

Scripture and Science Glorify the Creator

Total reality merges the material and spiritual worlds. Grasping the relationships is difficult, but possibilities present themselves.

Based on Ps.104:2, some creationists propose expanding of space after creation to explain star-distance enigmas. Heavenly-body distances might increase after creation, but Ps.104:2 Hebrew says God *stretchest out the heavens* (starry scene) *like a curtain*, as in stretching a curtain on a window (laying out or installing, not expanding). Extreme initial light speed can resolve the enigmas.

Creation possibilities: God's creation might invoke mass-energy equivalence ($E=mc^2$, c is light speed), making a tiny mass equivalent to a vast amount of energy to imply matter is frozen highly concentrated energy. Matter deriving from energy is observed by physicists at a sub-atomic level (most support this interpretation). Indeed it's implied by the inverse, loss of mass as energy in nuclear reactions (matter de-creation?) Another energy-matter link is a slight effect of gravity on light [46] paralleling the larger effect on matter. Energy and matter merge at a sub-atomic level.

God is the Light of the world with creation power in the sound of His *Let there be*. Application of this infinite energy would follow His conservation law, for energy wasn't created from nothing, being pre-existent in the Creator. This energy applied to creation would be conserved within Him and His universe together.

Accelerated isotope decay did occur, so if extreme radioactivity and heat implied by $E=mc^2$ at extreme light speed could occur, it was somehow negated for life to exist by the third creation day. Setterfield suggests energy-release prevention by conservation, light-speed change being countered by inverse rest-mass change.

Another possibility involves potential energy locked up in matter, substituting for initial active-energy release. Extreme initial light speed and isotope decay would be due to matter being in isotope form with extreme nuclei proton/neutron imbalance [47] that is extremely unstable, but is sustained by applied energy. As energy application ends, natural proton/neutron balancing would change atom properties to reduce light speed and isotope-decay rates in very rapid exponential fashion, and would produce non-isotopic and new isotopic matter. Resultant brief active-energy release would be applied to more matter creation, which would likely be followed by rapid attenuation of residual energy to control heat and radiation fully and thus ensure existence of life.

The brief active-energy release would create the hot earth core and overlying molten layer, and slowing isotope decay would produce heat now rising from the depths.* To support life and universe functions, consistent active energy continues as slowing solar/stellar radiation and isotope decay, and as interactive orbital gravity and relative motion unrelated to light speed (light speed has no effect on planetary orbits, as seen by freedom of dynamic clocks from the slowing atomic-clock time-keeping.

Exploding nova stars may arise from less-attenuated creation energy outside the solar system, due perhaps to star variant composition. Stars are giant nuclear furnaces, and accelerated isotope decay in the initial earth can reflect accelerated nuclear reactions outside our solar system. And the extreme matter-ejection energy of quasars suggests highly accelerated nuclear reactions. As conditions stabilize, explosions and ejection would cease, but appear at times varying with variant distance and decreasing light speed.

And a lesser isotope-decay acceleration could cause Great-Flood tectonic activity suggested by a Gen.7:11 breaking-up of fountains of the deep. Noah and his family would be shielded from radiation by deep flood water. They and the ark could be immune

---

*Early accelerated decay is suggested by correlation of earth natural heat flow with the amount of surface radioactivity, to suggest radioactivity heat dominates the flow. The heat isn't enough to yield observed flow patterns, if the current decay rates applied during the past few thousand years (Baumgardner, J. ICR. *Radioisotopes and the Age of the Earth*. p49).

to acceleration effects on radioactive elements in their bodies and the ark. Indeed a pre-flood status of their bodies, their food, the ark and the animals could isolate all these from earthly effects.

Thus Genesis 1, written ~1450 B.C, would present basic nuclear-physics concepts, revealing God's hand on the text. And perhaps Christ calmed a stormy sea, raised the dead and walked on water by speaking words invoking science laws that He alone applies, laws by which He spoke the universe into existence.

Closure: The Creator's nature determined the creation, so it was *very good* at first (Gen.1:31). Creation elements are Trinitarian in form, reflecting the nature of the Creator who is Father, Son and Holy Ghost. His people are Trinitarian, made in His image as a soul (controls, like the Father), body (Son) and spirit (Spirit). Water reflects His image, created as solid (Father, foundation), liquid (Son, the living Word by living water) and vapor (Spirit). And created first and second heavens are superintended by God's third heaven. And God's favored place in creation is Jerusalem, a present earthly one and a heavenly one superintending entrance into a final earthly New Jerusalem, where we who are His people shall be with God forever, and shall know even as we are known.

End Notes

1. Patterson, J.W. 1983. "Thermodynamics and Evolution". *Scientists Confront Creationism*. Ed. L.R. Godfrey. N.Y. W.W. Norton p99-115
2. Parker,G.1994. *Creation Facts of life*. Master Books Col.Sp. p110-118.
3. Parker. Op.Cit p155-165
4. Strahler, A.N. 1987. *Science & Earth History*. Prometheus. p489
5. Austin, S.A. Mar. 2000. Impact, Article 321. ICR. El Cajon, CA
6. Denton, M. 1986. *Evolution: A Theory in Crisis*. Adler & Adler. Bethesda, Md. Chap.12
7. Denton, Op. Cit. Ch. 5
8. Godfrey, L.R. 1983. "Creationism and Gaps in the Fossil Record." *Scientists Confront Creationism*. p193-214

9. Strahler, Op. Cit. p344

10. Morris, J.D. 1994. *The Young Earth*. Col. Spr. Master Books. p55,59

11. Morris, Op. Cit. p14

12. Carver, W. 1992. *The Science of Creation*. Bible Doct. Pub. p102

13. Ross, A.P. 2001. *Introducing Biblical Hebrew*. Baker. para.12.7.

14. Hebrew nouns said to be dual for no reason have one. Dual *Jerusalem* is a pair, the old city and New Jerusalem. Dual *Egpyt* is the nation and the world at large. *Waters* is a pair of creation waters, vapor above and liquid below the 1st firmament. Now *waters* for water bodies is plural in English convention, oceans, seas, rivers etc. relating to variant geographic locale, but *heaven*, dual in Hebrew, is often singular in English, for 1st and 2nd heavens are a pair in the form of one sky presenting the heavenly bodies.

15. Petersen, D.R. 1990. *Unlocking the Mysteries of Creation*. p26-29.

16. An ice-crystal shield atop the atmosphere is sustainable (icr.org/researchcanopy/pdf -*Temperature Profiles for an Optimized Water Vapor Canopy*, Vardiman, L). Small icy comets steadily enter the atmosphere so water as ice still exists atop the atmosphere (smallcomets.physics.uiowa.edu).

17. Humphries, D. R. *Creation Cosmologies Solve Spacecraft Mystery*. ICR. Acts & Facts. 10/07.

18. European. Space Agcy. Bul. 91. Aug.1997.

19. U. of Waterloo, ON. Press Rel.115. 7/17/97.

20. Morris, Op. Cit. 93-110

21. Mustain, A. An Amazing Planet contributor. LiveScience.com

22. Schweitzer, M.H. et al. *Science*. Vol 307, #5717. p1952-55. Mar.05 Follow-up work by Schweitzer and a large team confirmed soft tissue in "80 million-yr. old" dinosaur fossil bone dated by usual methods, refuting efforts of evolutionists to disprove soft tissue.

23. Brown, W.T. Ph.D. 1989. *In the Beginning*. Center for Scientific Creation. 5612 N. 20th Place. Phoenix, AZ. p15-19

24. Baumgardner, J. *Carbon dating Undercuts Evolution's Long Ages*. Impact #364. Oct.2003. Inst. for Creation Research. El Cajon, CA.

25. *Design and Origins*. 1983. ed. Mulfinger, G. Creation Research Soc. Monograph Series No.2.

26. Humphries, D.R., et al. 5th Int. Cong. on Creationism, Pgh. Aug. 2003.

27. Snelling, A.A. *Radiohalos-Significant and Exciting Research Results*.

Impact. Nov.2002. Inst. for Creation Research. El Cajon, CA.

28. Snelling, A.A. 2000. *Radioisotopes and the Age of the Earth.* p381.

29. Snelling, A.A. & Armitage, M.H. 2003. *A Tale of Three Granitic Plutons.* International Conference on Creationism.

30. Gheury de Bray, M.E.J. 1934. "The Velocity of Light." *Nature.* 24. Mar.1934. p464.

31. Bounds, V.E. 1990. "Further towards a Critical Examination of Setterfield's Hypothesis." *EN Tech. Journal.* 4:163-180.

32. Norman, T. & Setterfield, B. 1987. *The Atomic Constants, Light and Time.* Box 318. Blackwood, South Australia, 5051

33. Brown, Op. Cit. p89

34. Halliday, D. & Resnick, R. 1962. *Physics.* N.Y. Wiley. p1001

35. Chaffin, E.F. 2002. "A Determination of the Speed of Light in the 17th Century." *Design and Origins in Astronomy.* Vol. 2 Ed. DeYoung, D.B. & Williams, E.L. St. Joseph, MO. Creation Research Society Books.

36. Ratcliffe, Hilton. *The Jounal of Cosmology.* 2010. Vol.4. p693-718

37. Brown, Op. Cit. p90

38. Van Flandern, T.C. Precise measurements and Fundamental Constants 2. National bureau of Standards. Special Publication. 617. 1984. p625-27.

39. Troitskii, V.S. Physical Constants and the Evolution of the Universe. Astrophysics and Space Science. Vol.139. no.2 Dec. 1987. p389-411

40 Newcomb, S. *The Velocity Of Light.* Nature pp. 29-32, 13 May, 1886.

41. Dorsey, N. E. *The Velocity Of Light.* "Transactions of the American Philosophical Society," 34. (Part 1). pp. 1-110. October, 1944.

42. Birge, Raymond T. *The General Physical Constants.* "Reports On Progress In Physics." Vol. 8, pp.90-101, 1941.

43. Moffat, J. Inter. Journal of Modern Physics. D2, 1993. 351, 411

44. A. Albrecht and J. Magueijo, *Phys. Rev. D* 59:4 (1999), 3515; astro-ph/9811018, 2 November (1998).

45. J. D., Barrow, *New Scientist* 24 July (1999), p. 28.

46. A useful text for non-technical people on such matters is, Piccioni, R. L. Ph.D. *Everyone's Guide to Atoms, Einstein and the Universe*, but it's written from an evolutionist viewpoint that should be ignored, only the purely technical information being helpful.

Further Technical/theological interpretation for interested readers

<u>47</u>. Mass is inherent to atom-nuclei behavior, and $E=mc^2$ applies to isotope decay. Great heat/radioactivity of an earthly high-energy state might be prevented by compensating conversion of active (kinetic) energy to stored potential energy in matter creation, conserving total energy (the potential energy stored in matter replaces extreme active-energy release). That is, as creation energy is applied, isotopic matter would exhibit a kind of potential-energy supercharging by great temporal neutron/proton imbalance in atom nuclei. Resultant great instability would cause rapid deterioration of atom properties as energy application ends, so some "constants" would change rapidly in a rapidly-maturing earth. Such isotopes would decay exponentially, much faster than radioisotopes do today (reflecting decay differences like the extremely rapid one of polonium intermediate and the extremely slow one of uranium-238). Brief release of great nuclei binding energy could be applied to continuing matter creation to help minimize heat and radiation. Setterfield says heat/radiation impact could be like that of today due to invariant energy flow rate (ldolphin.org/cdkconseq.html).

Extreme light speed of an initial high-energy state requires extremes of radiation frequency and electron-orbital velocity, requiring extreme nucleus electrostatic charge and binding energy. This mode of light-speed acceleration seems possible, by temporal great neutron/proton imbalance associated with nuclear reactions creating starlight. Proving it would require evidence of extreme speed of early light from distant galaxies.

From a theological perspective, great apparent earth age caused by accelerated isotope decay would signify tempering of man's initial prospect of eternal life, warning of sin's curse as great apparent earth age forecasts all-encompassing death. It would be as if, in creation, God pictured a choice humanity would make between paradise and life as it is today. A potential holocaust of energy release by initial extreme light speed, if $E=mc^2$ applied initially, would signify the extreme horror of choosing sin, and if mankind did so, substitution of isotope supercharging for energy release would signify the Savior taking our judgment holocaust upon Himself to save us. Attenuation of light speed after supercharging would signify a majority of souls remaining in a state of slowing response to God's light.

The Creator's handiwork presents another explanation for extreme initial light speed. Light has properties of waves and particles. It's energy, but its very slight reaction to gravity and a slight pressure it exerts on tiny bits of matter (www.answers.com/topic/poynting-robertson-effect), are physical behaviors very slightly like that of mass, accounting for decreasing light speed by interaction with dispersed matter in space. The decrease is exponential, reflecting usual energy behavior, so this aspect of its electromagnetic energy-type behavior isn't affected by its very slight mass-like behavior. Thus the latter just mimics true mass behavior to a degree.

Now we consider gravity, the pull of which is instantaneous, as studies have indicated (www.metaresearch.org/cosmology/speed_of_gravity.asp). A claim that speeds of gravity and light are the same (www.newscientist.com/article/dn3232) seems just meant to retain universality of $E=mc^2$, a type of behavior common today (www.lbl.gov-science-articles/archive/phys-speed-of-gravity.html). Gravity must apply instantaneously, or bodies in orbital relationship will spiral into each other (Eddington, A.E. *Space, Time and Gravitation*. 1920. Reprint, Cambridge Press. 1987. p94).

$E=mc^2$ doesn't apply to gravity and so isn't universal, and instantaneous gravity pull prevents dire effects on orbiting bodies, as supported by freedom of dynamic clocks from the slowing time-keeping effect related to light. Contrasting behavior of gravity and light offers a clue to explain extreme initial light speed. If creation invoked an ideal environment with light travel like that of gravity, instantaneous, due perhaps to a fabric-like field in space (like the *ether?*) propagating an instantaneous tensile-like pull in the case of gravity or an instantaneous wave-like photon continuum in the case of light, $E=mc^2$ wouldn't apply, and extreme light speed would not release extreme energy. Light would be free of mass-like effects, and current photon nature reflecting energy, but mimicking mass nature very slightly, would result from a change in the initial nature.

Theologically speaking, in creation, instantaneous gravity extending from the furthest reaches of heaven would signify an instantaneous link to God and the prospect of eternal life. Decreasing light speed may be due to light losing the nature of pure energy waves, signifying loss of purity of God's light in human souls created innocent. This would reflect sin's mass/weight that diminishes human response to the light of God's Living and written Word, to a level far below that offered initially, yet still available and steadily decreasing by just tiny amounts with the passage of time. Gravity retains a pure nature, still linking us to God by uncontaminated vast reaches of the heavens, and signifying salvation still possible for all who desire purity from God over the mass/weight of sin's defilement.

Creation pictures mankind, made in the image of God, at the center of His purposes in a creation vast and complex beyond our comprehension.

The KJB in Relation to Inspiration, Inerrancy and Providential Guidance

By Dr. Larry Bednar

Foreword

The issue of Bible translation is complex. Those of us who believe by faith that God kept His promise to preserve His Word, grapple with questions of inspiration, inerrancy and preservation. Dr. Larry Bednar does an outstanding job in presenting providential guidance as a major aspect of the issue, illustrating undeniable KJV accuracy in cases that lead us to believe God had something to do with the translation. In the academic world, some who follow lexicons reject the KJV as their standard. We, however, believe the KJV is the best lexicon for Bible translation, and this fact is very well presented in this treatise. We are pleased to have Dr. Bednar as a member of our staff, and we acknowledge the views of this treatise as our own.

Daniel S. Haifley, Th.D.,D.D.
President of Indiana Fundamental Bible College

Preface

From topic 1, readers can see that evolution completely fails to explain the existence of life and the character of the world we live in, and the only alternative is creation by God. Here in topic 2, we'll find that the Christian Bible shows evidence of being God's revelation and thus is the one source of information about the creation and all else that God would teach to all whose faith is in Him. The accuracy of that word in our traditional scriptures is indicative of text inerrancy preserved from inspired autograph originals and extending to our English translation, the traditional KJB. The result is great confidence, not only in knowledge of the creation, but all matters of life and eternity.

Introduction

Scholars say scripture inerrancy is limited to long-lost originals, making it unverifiable. But why would only some earliest readers of Hebrew/Aramaic or Greek have inerrant texts? Isn't everyone entitled to inerrancy if all are to live and be judged by God's word? (Mt.4:4, Jn.12:48 - error can render judgment unjust). That God would inspire inerrant autographs just to let inerrancy be lost, isn't credible, full accuracy being vital to His people. Logically, He will guide select copying and translation that, in finished form, are equivalent in effect to inspiration, despite basic differences in the methods. Thus His progressively-revealed inerrant word would be preserved for His people at their times in history.[1]

But scholars invent critical Greek texts, choosing readings they prefer in rediscovered long-lost manuscripts (they see themselves as the means of preservation). They deny full accuracy of copies or translations and assume their preferences are fine. We refute this, making a case that God's Word can only be traditional texts, fully preserved by Him and revealing His hand on them, that His people may know His will with certainty. We begin making our case with evidence of God's hand linking inspiration to select translation. The KJB shows this evidence by unique consistent accuracy indicative of inerrancy, in turn indicative of an inerrant textual basis reproducing inspired inerrant autograph originals. Accuracy is striking in passages where scholars deny it.

Defining and Illustrating the Autograph-Inspiration Method

Plenary/verbal inspiration requires that each word be dictated to men's minds by the Spirit to achieve inerrancy. Scholars dismiss dictation due to human input in the form of writing style and supposed error in scripture. Consistent KJB accuracy defies the notion of error, and human writing style is seen since, contrary to how scholars define it, dictation doesn't suspend writer intellect.

---

[1]. L. Bednar. *The Guide to our Eternal Destiny*, Appendix D. WCBS/ IFBC book: God's Word in the Life of His people: The Available History.

Dictation takes variant form. Writing style can be absent due to verbatim dictation, as in Exodus 34:27 that says, *And the Lord said unto Moses, <u>Write thou these words</u>*...Jeremiah 30:1,2 says, *The word that came to Jeremiah from the Lord, saying, Thus speaketh the Lord God of Israel, saying, <u>Write thee all the words</u> that I have spoken to thee in a book.*

A vision dictates to a writer's eye and ear. Revelation 1:10, 11 says...*I was in the Spirit on the Lord's day, and heard behind me a great voice, as of a trumpet, saying...<u>What thou seeest, write</u> in a book*...Dictation is also to the ear. In 2:1 Christ in the vision says, *Unto the angel of the church of Ephesus <u>write</u>; <u>These things saith he</u>*...Intact faculties are indicated. In 1:10 John hears a voice behind him and turns and gives details of what he sees. In 1:17 he falls at the feet of Christ in the vision, as expected of one with intact faculties confronted by a representation of God.

Dr. Phil Stringer, Ravenswood Baptist Church, Chicago, notes scripture's most unique case of dictation, the speech of Balaam's donkey. God spoke to Balaam by the animal in words recorded in Num.22:28, *And the Lord opened the mouth of the ass, and <u>she said</u> unto Balaam, What have I done unto thee, that thou hast smitten me these three times?* The ass had no possible speaking style, yet seems intellectually involved, an impossibility, so God preserved her "viewpoint." Thus preservation of viewpoints, and so mental faculties, of scripture penmen is expected.

Dictation can be so unique a writer exercises normal intellect and style as Holy Ghost supervision of his writing by-passes his awareness (God's still small voice). Scripture reveals dictation of acts and words of people without their knowledge. We consider Jesus' mother Mary, *great with child* (Lk.2:5) in Nazareth ~75 miles from Bethlehem. If she or Joseph knew the Micah 5:2 prophecy on a Bethlehem birth, in the hardship of the final stage of pregnancy, she'd dread a difficult trip to Bethlehem. And Joseph would dread hazards of labor pains, miscarriage or premature delivery in a lonely area on the way. God had to handle such details, but how would He get her to Bethlehem? He'd stir the mind of a pagan, Caesar Augustus, to see a need to finance the empire in ways better than resented harsh taxes (Lk.2:1-5). There likely was a fear of tax rebellion then (Acts 5:37 tells of tax rebellion when the Roman governor of Syria was Cyrenius,[2] who ruled earlier when Mary and Joseph went to Bethlehem - Lk.2:2). God would move Caesar to require empire residents to enroll for taxation at their native cities to establish connections of each to family assets through local authorities (an old private-contractor system using locals would be fair, but publicans like Zaccheus would cheat). All would pay by ability, the masses not being affected enough to incite rebellion (it occurred later). Augustus introduced a graduated income/property tax (Thompson finds the tax enrollment began with Augustus,[3] as is logical, for it accords with his *Pax Romana*, a time of relative peace, prosperity and tax reform). The taxation summons was timed too close to the birth of the Child to allow delivery before Mary and Joseph had to go to Bethlehem. Thus they came, in timely fashion, to their city of origin as descendants of David, because God subtly dictated the acts of many people in a vast empire without anyone realizing it.

Regarding word dictation, in Ps.22:8 David speaks words of men who persecute and taunt him, saying, **He trusted on the Lord that he would deliver him: let him deliver him**...This prophesies words spoken by Jesus' persecutors at the Cross 1000 years later. In Matthew 27:43, priests, elders and scribes taunt Jesus saying, **He trusted in God; let him deliver him**. The prophecy is fulfilled on behalf of Jesus by His worst enemies, authenticating Him as the Son of David, God's Messiah, the last thing in the world they would willingly do. Had they known this meaning, they would never have spoken these words. Thus dictation by-passes speaker awareness, and makes use of malicious free-will word choice, literal word meaning and speaking style. It's said they mocked

[2]. *King James Bible Commentary*. 1999. Nashville. Nelson. p1343.
[3]. Thompson, J.A. 1962. *The Bible and Archaeology*. Eerdmans. p375.

Jesus with David's prophecy, but they revered David and would not identify themselves with men vilifying him. And they would never relate prophetic words of David to Jesus, lest <u>they</u> present prophecy authenticating Jesus as the Messiah. Actually they'd not see the verse as prophecy useful in feigned fulfillment, for, of itself, it's indicative only of David's history, not of prophecy.[4]

Now would the Spirit work in men just used by satan? In Mt. 27:42 they said of Jesus...*let him now come down from the Cross, and we will believe him*, opposing salvation. The one alternative to a Spirit role is God putting just the right men of just the right free will in just the right place to fulfill His will in just the right way. With the prophecy made ~1000 years before the Cross, this would mean God logistically or genetically controlled ~40-50 generations for 1000 years, intervening in their decisions to establish His will. Mt.10:29,30 supports, saying, *Are not two sparrows sold for a farthing? And one of them shall not fall on the ground without*

---

[4]. Scholars say the initial Ps.22:8 clause should be the imperative, <u>Commit yourself to the Lord</u> (NASV), not <u>He trusted on the Lord</u>, which would nullify Mt.27:43 fulfillment of the Ps. prophecy. Hebrew grammar indicates an imperative or infinitive, but an imperative would link a 2nd person command (*Commit yourself to the Lord*) directly to a 3rd person declaration (;*let Him deliver him*) in one thought in absurd grammar/syntax; the scholars justify this, saying the clause after the imperative is an aside, with taunters talking to each other. Well, declarative clauses with a difference in person can link in poetic/prophetic Hebrew to vary perspective (e.g. Is.52:14, 61:7). But linking a command to a declaration, with a difference in person, all in one thought, requires extra words not in the text.

The infinitive expressed as perfect tense (*He trusted*) is indicated by the inerrant New Testament. Ps.22:8 relates to Mt.27:43 priests and elders who address the people, but aim their talk at Jesus on the Cross to taunt Him (indirect taunting is differentiated from Mt.27:40,44 direct taunting of others). Words of indirect address aimed at Jesus but spoken to others, fulfill prophecy on earlier indirect address aimed at David but spoken to others. The imperative applies only to direct address (e.g.Ps.37:5, Pr.16:3), and there's no verb form for the indirect. The form needed will correlate with jussive verb sense (wish) concluding the verse, as the infinitive expressed as perfect-tense does (*He trusted in the Lord...let him deliver him*).

your Father. But the very hairs of your head are all numbered.

If the Spirit did not speak through these men, dictation was indirect, through control of many generations. Otherwise it was direct, using minds and speaking styles of evil men against their will, with no appearance of dictation. One way or another, God made evil men in control of their faculties speak His words.

Thus the Spirit dictates to obedient servants, unknown to them, which would be by thought motivation. Luke 1:3 says, *It seemed good to me also, having had perfect understanding of all things from the very first, to write...* This gospel seems written at a whim from memory, but the Spirit directing Luke without his awareness explains *It <u>seemed good</u>...to write,* which explains his <u>perfect</u> *understanding...* Now we see the passage very differently. Luke writes in his style by his intellect, and the Spirit motivates his free will, supervising the writing by allowing or disallowing word choice (i.e. keep this word Luke, not that one). God dictates each word, editing words produced by Luke's intellect and style to avoid all appearance of dictation. 2 Peter 1:20,21 suggests such inspiration, saying, *the prophecy came not in old time by the will of man: but holy men of God spake as they were moved by the Holy Ghost* (men's words as willed by the Spirit).

In Psalm 69 David tells of his reproach and persecution, relating his words to those of the Son of David on the Cross 1000 years later. David's words begin resembling those of Jesus at the Cross in verse 7, and on reaching verse 21, he speaks Jesus' own words of His persecution and reproach and toxic gall and pain-numbing vinegar offered to Him (Mt.27:34). A transition from David's words to those of Jesus is like parting of a curtain to reveal the mind of the deity dictating all of David's words. David speaks in regular human terms of himself, but his words that are Jesus' literal words relate only figuratively to himself and are those he wouldn't normally say of himself, further indicating dictation.

A 3-fold dictation, in a prophecy on betrayal of Jesus for 30 pieces of silver, appears in Zechariah 11:12,13 that says, *So they weighed for my price thirty pieces of silver. And the LORD said unto me, Cast it unto the potter...And I took the thirty pieces of silver, and cast them to the potter in the house of the Lord.* The speaker of *my price* can only be Christ pre-incarnate speaking by (dictating to) Zechariah. Judas, after the betrayal, is moved in his mind (dictated to) by Christ to do the right thing, to return the blood money to the priests and elders; he casts it down in the temple in disgust at what he's done, selling his soul for money. And priests and elders causing the Crucifixion, normally covetous men, used the money for the potter's field to bury strangers (Gentiles they hated), being moved in their minds (dictated to) by Christ to use the money properly. Christ figuratively cast the silver to the potter in casting it down in the house of the Lord, doing so by dictating thoughts of Judas and priests and elders.

Equivalence of God-Guided Translation and Inspiration

Objective textual study shows the KJB is consistently accurate, and accuracy is striking in passages where scholars vehemently contest it. Indeed KJB accuracy is so consistent it appears to be total and thus indicative of inerrancy that, in turn, is indicative of an inerrant textual basis reproducing inerrant autograph originals. And such English-text accuracy is realized despite very different grammar of underlying Hebrew/Aramaic and Greek languages.

All this points to providential guidance of an inerrant-text history concluding in the KJB. Inerrancy will identify the KJB as a written-word offspring of virgin inspiration in the true textual family, even though inspiration differs greatly from hard work of translation scholarship. Translation ordained and guided by God would be equivalent in effect to inspiration. The Spirit dictated autographs undetected, and He can guide translator thought or arrange textual matters undetected, with full equivalence. The equivalence will continue as God ordains new editions with new language convention for new generations, but modern English versions are excluded, being just man's work utilizing humanist scholarship and deficient language unworthy of the sacred text.

Full translation accuracy, vital to many who can't read Greek or Hebrew, necessitates inerrancy, which is attained only by providential guidance. KJB consistent accuracy indicative of inerrancy is illustrated below, and in other treatises of this writer (ref.14). Scholars are to teach recognition of such matters, but their anti-KJB humanist scholarship misleads them and their followers.

1. An ultimately-unique word sense: *Replenish*, Gen.1:28.

An evidence of KJB providential guidance in the Old Testament would be prophecy fulfillment seen only in the KJB, while Hebrew-text inerrancy is maintained, which occurs in Genesis. Some evolutionists propose a millions-of-years gap from Gen.1:1 to 1:2 in which man-like creatures lived and perished, saying the KJB 1:28 *replenish the earth* means Adam was to repopulate earth. But *replenish* meant *fill* in 1611 England (*replete* still means *filled*), and the translators didn't know of the future *refill* sense. Use in a refill sense occurred sporadically after KJB publication (Oxford Eng. Dict), and was not established until over 200 years later, still meaning *fill* in Webster's 1828 dictionary.[5]

Use of *replenish* in Genesis is unique. With *refill* applying long after 1611, the 1:28 command to replenish can be seen as signifying veiled prophecy on earth's population loss in the great flood, *replenish* appearing in the same command in Gen.9:1 after the flood to fulfill 1:28 prophecy. That is, God told Adam to fill the earth, and prophesied to Noah in Adam's loins to refill it after the flood (as in Heb.7:10: Levi in Abraham's loins, centuries before his birth, pays tithes to Melchisedec). The Hebrew word means *fill*, but can be rendered *refill* if that's the contextual sense of *fill*.

KJB translators unknowingly linked *replenish* in Genesis 9:1 and 1:28, making these the only uses in Genesis and linking the creation to the great flood.[6] They rendered Hebrew for *replenish*,

---

5. Daniels, D.W. 2002. Chick Publications web site on Bible versions.
6. The water-vapor canopy placed above the firmament in creation is a source of 40-day rain of the flood: God foretells the flood at creation.

fill in a 1:22 command in the same context as 1:28, regarding fish not fully destroyed in the flood and <u>not in need of refilling</u>. This Hebrew is used in like sense in 7 other Genesis verses, and they rendered it *fill*, not *replenish*, and *refill* doesn't apply in any of these verses. How can this be when they didn't differentiate *fill* from *replenish*? It appears God guided them by a need to replace antiquated *plenish* (fill) with *replenish*, English versions consulted providentially placing *plenish* where *replenish* would substitute later, and much later would mean *refill* (God's work, for no version writer knew the future). KJB translators would be aware only of a textual item, but 9:1 fulfilling of 1:28 prophecy, <u>visible only in the KJB and long after 1611</u>, reveals God's guidance in arranging a textual item. *Replenish*, marking God's hand on the text, must be retained, and Blayney's 1769 edition providentially ended KJB language up-dating that alters words of this type.[7]

2. Christ's close scriptural-ministry tie to John the Baptist

Details of God's plan revealed in the New Testament only in the KJB, while Greek-text inerrancy is retained, will reveal KJB providential guidance, and this occurs in Mt.14 and Lk.7. Evidently God signified the end of the Old Testament era in a way that gives precedence to the New, as indicated by criticism of the KJB for making John's baptizer role a last name. Lk.7:20 says *John Baptist*; Mt.14:8 says *John Baptist's head*...Now to say *John the Baptist* can't be called *John Baptist* is to say *Jesus the Christ* can't be called *Jesus Christ*. *Christ* and *Baptist* signify roles that become names.[8] *Jesus Christ* is equivalent to *Jesus the Christ*, as is true of *John Baptist* and *John the Baptist*; either name can be rendered.

In both verses the Greek reads *John the Baptist*, as the KJB usually has it, but can be rendered the equivalent *John Baptist* if

---

[7] By this Providence, other older words that can be up-dated aren't, indicating suitability of older language and unsuitability of ever-degenerating modern language in the scripture text (glossaries can define older words).
[8] More examples: *Simon Zelotes* signifies *Simon the zealot*, avid follower & *Mary Magdalene* signifies *Mary the Magdalene*, resident of Magdala.

context specifies it. It appears in two gospels and the context is the same in both, John's imprisonment and death (He's in prison in Mt.11:2 and thus during the discourse of his disciples and Christ in Lk.7:20), so the matter is one of deliberate design.

Context specifies *John Baptist* as a proper alternative to John the Baptist in the two verses. Christ and John brought in the New Testament era (John prepared the way, introducing Christ by a miraculous birth partly reflecting the Virgin Birth, and by baptism of Jesus in Jordan, and he preached and baptized for repentance vital to salvation). John's work was soon to end at the time of the Mt.14 / Lk.7 event; as the last Old Testament prophet, his death ended Old Testament ties to the work. *Jesus Christ* is Jesus the Christ, anointed Savior whose death and resurrection began the New Testament era. And *John Baptist* is John the baptizer, the role of the last prophet whose death ended the Old Testament era (he said of Christ, *He must increase, but I must decrease* - Jn.3: 30). John is a unique Old Testament prophet, for while Moses, Isaiah and others foretold Christ, John introduced Him. *John Baptist* identifies him, by the role, with *Jesus Christ* in bringing in the New Testament era (they were even kinsmen by the earthly tie). The tie is also seen in that, as Jesus had God's Spirit beyond measure (Jn.3:34), John was filled with God's Spirit before birth (Lk.1:15). And the Mal.4:2 *Sun of righteousness* with healing in his wings is Christ, while the Mal.4:5 *Elijah* introducing the day of the Lord is John in association with Christ [1st & 2nd Advents are noted in Malachi, with Elijah-type John at the First Advent (Mt.17:9-12) and the literal Elijah at the Second (Rev.11:3)].

A *John Baptist* tie to *Jesus Christ* signifies the end of the Old Testament era at John's death, giving precedence to the New soon to begin by Jesus' death and resurrection. Death ends the Old Testament era, but Resurrection victory over death begins the better New era. God's guidance is KJB-translator recognition of veiled contextual sense controlling the rendering to tie John's Old Testament status to his New Testament baptizer role. John's forerunner baptism of repentance in death to self-desire, the Old-Testament final one, is superseded by Christ's baptism of the Spirit (Ro.6:4) signifying death to the entire old man and resurrection to new life. This reflects John's death and superseding of the Old Testament era by Christ and Resurrection life of the New Testament era.

3. Unique word sense from context/history: Easter, Acts 12:4
Another evidence of providential guidance is translation scholarship that can't be matched, the greatest scholars being selected for work ordained by God. Great scholarship includes knowledge far beyond the textual, in history and other humanities, to permit recognizing contexts requiring unique renderings; this is the case for a much-criticized KJB Acts 12 rendering of *pascha* as *Easter*.

12:2 *And he* (Herod) *killed James the brother of John...*
12:3 *And because he saw it pleased the Jews, he proceeded further to take Peter also. (Then were the days of unleavened bread).*
12:4 *And when he had apprehended him, he put him in prison ...intending after <u>Easter</u> to bring him forth to the people.*

Pascha is usually *passover* in the KJB, but Providence preserves something unique in this context by the KJB. The case is like that of the Isaiah 7:14 *virgin* regarding Mary and the Lord's Virgin Birth. The Hebrew for *virgin* can mean *young woman* or *maiden*, but Isaiah 7:14 context and related word sense demand *virgin*. *Pascha* has more than one possible meaning, and *Resurrection Day*, or *Easter*, is demanded by history and Acts 12 context.[8]

In 12:2,3 Herod killed James and imprisoned Peter in the *days of unleavened bread*. Others have noted this can refer to the feast of unleavened bread, 6 days following Jewish Passover day [in Lev.23:5-6 & Ex.12:18 Passover is at evening on the 14th day of

---

[8]. Early English versions first use *Easter* generally in the N.T, but later favor *Passover*. The Bishops' Bible retains *Easter* at Acts 12:4, but makes *Jews' Passover* an untenable *Jews' Easter* at Jn.11:55, indicating a lack of analysis of *Easter* usage. The KJB rightly limits *Easter* to Acts 12:4.

the proper month, and the feast of unleavened bread is 6 days from the 15th (at evening) until the 21st day (at evening) – Including Passover day yields 7 days of unleavened bread of Lev.23: 6]. If Herod killed James and imprisoned Peter in the 6 days of unleavened bread, Jewish Passover day was over, and an Acts 12 later Passover wouldn't refer to normal Jewish Passover day.[9]

Yet Acts 12 can refer to Passover day plus the next 6 days of unleavened bread. Ezekiel 45:21 calls the term a 7-day event, saying...*in the fourteenth day of the month, ye shall have the passover, a feast of seven days; unleavened bread shall be eaten.* This accords with Exodus 12:18 that tells of eating unleavened bread in the evening from the 14th until the 21st day of the month (7 days of unleavened bread), and Leviticus 23:5 that says the Passover meal is on the 14th day at evening. A 7-day passover can change Acts 12:4 interpretation. It might seem that Herod killed James and imprisoned Peter during the combined 7 days of Passover day and unleavened bread, which hadn't ended at this point, and he might plan to bring out Peter at the end of this 7-day Jewish Passover. But this interpretation proves to be wrong.

Context/history deny *Passover* in Acts 12:4, which could only be Jewish Passover in this context on days of unleavened bread, and that doesn't fit context/history. Lack of a fit relates to Herod's friendship with Roman Caesar Caligula (hated by the Jews) and his ancestry as an Edomite, a people historically antagonistic to Israel. Herod's rule was difficult,[10] but he endured, ingratiating himself with the Jews by favoring their religion and culture.

Acts 12:3 says Herod took Peter prisoner since James' death pleased the Jews. They'd see James as an enemy as fast-growing Christianity threatened their religion and way of life. Thus there was no problem with the Jews in killing Peter right after James, and Herod had no reason to wait until after Jewish Passover, the first day or the 7 days. Yet he would wait, risking a problem with

[9]. Moorman, Jack. *Conies, Brass and Easter.* Way of Life literature.
[10]. Broadman Bible Commentary. Vol.10. 1970. Nashville. p75-76.

the Jews by a suggestion of changing his mind in acting against the foremost leader of the fast-growing church.

And scholars are wrong to say Herod had to wait until after Jewish Passover to kill Peter since Jews objected to executions during their holy days. That was usually so, but he had no reason to worry about that <u>at this time</u> in history. Christians were now considered heretics by the Jews, so a public execution reinforced the Jewish position. As Acts 12:3 says, the Jews approved of the killing of James during holy days of unleavened bread at that time. Thus context and history deny interpreting Acts 12:3 to mean Herod would wait until after Jewish Passover to kill Peter.

But Pascha as Resurrection Day fits context/history. The Jewish 1st- century church knew Jewish Passover was typological, being fulfilled and superseded by Christ the ultimate Passover, so their Pascha would be based on the Crucifixion/Resurrection. Timing of this Pascha would fit Acts 12 days-of-unleavened-bread context/history since the only initial timing basis was Crucifixion/Resurrection timing relative to that of Passover. The Crucifixion was on a preparation day for Passover starting that evening (Mt. 27:62, Jn.19:14), requiring a 3-day observance at the same time relative to 7-day Passover, both events starting at evening for a close relationship.[11] Herod could kill James and imprison Peter on an evening that began a Jewish Passover day, or on the next day, which would be during the days of unleavened bread, and he might wait to kill Peter until after the third day, Resurrection Day. Herod's political situation indicates that this was the case.

Herod was accepted by the Jews, but despised by Judean

---

[11]. 1st-century church pascha was a Crucifixion/Resurrection observance starting on the evening Jewish Passover began (no. of days & other details are murky). In the 2nd-century eastern church, a 1-day Pascha at the same time likely reflects 1st-century timing of a 3-day Pascha starting that day. In starting pascha observance on Crucifixion Day and making it a 1-day event, Resurrection Day is observed on a day signifying Crucifixion Day, an irregularity likely due to an earlier 3-day event starting the same time (*Easter* and *Paschal Controversies*. "Evang. Dict. of Theo." 1984. Baker).

Romans (Lk.23:12)[12] crucial to his rule, causing political tension. Pleasing the Jews was a priority, but his Roman trouble required avoiding antagonism of any large segment of Judea's population that could cause political unrest and give local Roman leadership an excuse to depose him (a good excuse was vital as Caesar appointed Herod). He'd worry over Christian reaction to killing Peter (The sizeable Christian population still had political status, for Roman rulers wouldn't persecute them for another 20 years). He killed James without repercussion, but would worry that killing Christ's most famous disciple, Peter, right after James, might incite an uproar, especially if he did so at Resurrection-Day time. To Christians this day is sacred, and public killing of two great disciples at this time would be very politically antagonistic, mocking Christ's victory over death. In this act Herod would make a political statement like, "Is this your day of victory of eternal life over death? I'll kill Christ's great disciples at this time to make it a day of death. I'll show you what I think of your God." While Christians endured persecution, Herod knew killing famous Peter right after James, and insulting Christ, might incite an uproar. By executing Peter soon after Resurrection Day, he'd placate the Jews without unduly risking great Christian protest adding to his Roman problem. This would seem wise to Herod and is why he would wait until after Christian Passover, not the Jewish one.

Pascha in Acts 12 refers to Christ, Passover of Christians (1 Cor.5: 7). For Jews Passover observance was 7 days as Ezekiel 45:21 says. But 1st-century Christian-Passover observance would be 3 days from Good Friday to Easter. Herod planned to execute Peter after Christian Passover to avoid antagonizing Christians, or more specifically, after Resurrection Day. This very important day for Christians would be the day of greatest concern to Herod.

Actually, in the Resurrection, Christian Passover superseded the Jewish, so *pascha* is Christian Passover after the Resurrection, or after Acts 1 in scripture, where it appears 3 times. A Hebrews 11:

12. Funk & Wagnalls Encyclopedia. Vol.13. p81.

28 use refers to Old Testament times and isn't pertinent. The 1 Cor.5:7 use is rendered *Passover*, for this is the verse where Christian Passover supersedes the Jewish, and it clearly is the Christian one since Christ and the Cross are its basis. Acts 12:4 use can only refer to Christian Passover. *After Easter*, more specific and better understood by readers, replaces *after Christian Passover*. *Easter* communicates the sense of Resurrection Day, despite a 1st-century observance different from that of later times.

Some reject the term *Easter* since it didn't have a Christian sense until long after writing of the New Testament, making its use in Acts 12:4 anachronistic, out of place in time. But Pascha was Resurrection Day specifically by the 4th/5th century, and it was Easter by the 8th. Since the 8th century, *after Easter* has been the right current way to note the end of the 3-day Passover Herod had to wait out to kill Peter. So why are 17th-century KJB translators scorned for proper use of current terminology? Modern translators often do so and are praised for communicating with moderns. For example an NIV *gallons* (Lk.16:6, Jn.2:6) is fine today, but it's out of place in time, and out of place in culture, having never applied to Hebrew, Greek or Roman culture, so *gallons* is further removed from a 1st-century setting than *Easter*.

The KJB is the one English translation affirming 1 Cor.5:7 teaching on Christian Passover superseding the old. The KJB precisely reflects the Greek text, using *Easter* to signify the last day of Pascha to avoid uncertainty on the number of days Pascha lasted in the 1st century and to specify the day Herod had to get past. But today scholars treat translation precision as if it were error! Providential guidance resulted in selection of KJB scholars with skill and insight far exceeding that of modern scholars. Guidance would include directing scholar history knowledge to pertinent contextual factors during translator-group debates on renderings.

4. The "modern" science principle of wind generation.
Another indication of KJB providential guidance is preservation

of modern-science truth in an ancient Bible book written ages before men knew this truth, and translated long before they knew it. This is the case with wind-generation technology noted in Job, written ~4000 years ago (contrary to scholars - see appendix).

Job 38:24 God's question for Job.

KJB: *By what way is the light parted, which scattereth the east wind upon the earth?*

NASV: *Where is the way that the light is divided, Or the east wind scattered on the earth?*

NIV: *What is the way to the place where the lightning is dispersed, or the place where the east winds are scattered over the earth?*

Wind arises as solar energy heats earth's surface, causing heating and rising of air and lateral air flow to fill a void that would be left by rising air. The ancient Job book reveals this "modern" science concept, saying *light scatters* (moves) *the east wind upon the earth. Light* is correct, conversion of light to heat on absorption by earth's surface being the heat causing wind (conversion of light to heat is seen by burning paper with a lens focusing the sun's light, not its heat – reflects the 1st law of thermodynamics formulated in 1842). The ancient writer couldn't know this technology, so it's due to providential leading. The concept would be hidden from the ancients due to night winds from temperature differences like those prevailing between land and sea after daytime solar heating. Thus the KJB tells us the Hebrew text reveals God as the ultimate writer here, and this must be preserved.

Job gives the basics, and also further technical detail, on the wind-generation mechanism. The clause *by what way is the light parted*, reveals concisely how wind is generated. *By what way*, in reference to *east wind* (long-distance wind), notes the way in terms of the manner and direction involved as sunlight is *parted* (intensity is partitioned or divided) to move air and generate winds of different specific force. Sunlight strikes earth at various angles from the vertical at various longitudes and latitudes due to the earth's surface curvature, and earth heating varies with the angle of inclination of solar radiation. Due to parting of light intensity with the angle of inclination, variant temperatures arise on earth's surface, creating wind cells in which air moves from a cooler higher-pressure area to a warmer lower-pressure one.

Thus, to the question *By what way is the light parted*, the answer in terms of manner is, by earth's surface curvature, which Job alludes to. The answer in terms of direction is, along curved earth paths on which variant angular directionality of sunlight varies earth heating in adjacent regions to produce wind cells of specific variant ranges of force and direction on a rotating earth (direction is illustrated by <u>east</u> wind, an east-to-west direction). This modern-science knowledge concisely revealed in an ancient bible book can only be due to the Creator's omniscience.

The light/wind connection in this ancient book proves the divine hand on the text, and modern translators <u>break this connection, removing evidence of the divine hand</u>. Hebrew grammar makes *parted* passive voice (light is acted on, it *is parted*) and *scattereth* active voice (light acts on, it *scattereth* wind).[13] The result is two roles of light in one thought. Modern translators alter grammar to make sense of the verse when missing the light/wind connection. They make both verbs passive, separating the light roles into two thoughts using *or* for *which*. Neither word is in the Hebrew, and grammar/context determining the choice indicates *which/that*.

It's as if modern scholars erased God's signature from the text, distorting Hebrew grammar and altering verse meaning because they didn't know some well-published technology. Despite lack of knowledge of the technology at their time in history, KJB scholars preserved it by honoring the grammar, thus preserving evidence of God's hand on the text. They likely doubted a light/wind connection due to simpler 17th-century science, but Spirit guidance would stir-up in these reverent minds, a reverence for

---

[13]. Hebrew for *parted* is niphal (passive) & for *scattereth* is hiphil (active).

God's Word that overruled men's opinions and knowledge.

More poor science is seen in the NIV *dispersed* and NKJV *diffused* indicating light scattering. Orderly light parting is what creates winds of specific force, and wind, not light, is scattered.

5. For whose sin did Christ suffer? 1 Peter 4:1,2
Another evidence of providential guidance is avoiding Greek or Hebrew texts distorted by tampering in underlying manuscripts. Often Greek-manuscript tampering attacks Jesus Christ's deity.

KJV: *Forasmuch then as Christ hath suffered <u>for us</u> in the flesh, arm yourselves likewise with the same mind: for he that hath suffered in the flesh hath ceased from sin; That he should no longer live the rest of his time in the flesh to the lusts of men...*

NIV: *Therefore, since Christ suffered <u> ? </u> in his body, arm yourselves also with the same attitude, because he who has suffered in his body is done with sin. As a result he does not live the rest of his earthly life for evil human desires...*

The KJB teaches that Christ's suffering was for our sin, so we honor His great sacrifice. As Christ was willing to suffer for our sin, we should be willing to endure hardship God brings our way to kill appetites of the flesh. In this way we partake of Christ's sufferings on our behalf (1 Pet.4:12-16) that we might conform to His death (Philip.3:10). We begin to reject sin as it becomes ugly to us by the hardship it brings our way, reminding us of the suffering it brought to the sinless Christ on our behalf.

The NIV follows a Greek text that omits the crucial note that Christ suffered <u>for us</u>. Now when it's not specified that His suffering was for us (for our sin), the logical interpretation is that Christ suffered to be done with His sin and that we should follow the example and suffer to be done with ours. This would make Jesus a mere sinful man suffering to atone for His own sin, and us following His example and suffering to atone for our sin!

This omission, that attacks Jesus Christ's deity, and would

defeat biblical salvation teaching, is one of many such errors the KJB avoids. The Spirit would direct KJB translator reverence in selecting the textual basis, or making distorted texts <u>unavailable</u> (scholars say the Received Text is based on too few manuscripts, and they suggest utilizing Alexandrian-type manuscripts, the type with error like that seen above in the NIV 1 Peter 4:1,2).

Final note: Another evidence of providential KJB guidance is retention of true readings despite minor manuscript support, for a true reading can't be left out of God's true Word. This is the case with the Johannine Comma (1 Jn.5:7-8), authenticated by context and language, plus evidence of a connection to the autographs, plus association with what looks very much like God's plan for a new widened true-text distribution in 16th-17th century Europe.

At times manuscript support relates to church size, and biblical churches that would fully preserve God's Word were small and few in medieval history, so copyists and copies would be few compared with those of large unbiblical churches. A few true readings meant especially for hearts of true bible-believers, and apt to be disdained by irreverent unsaved persons in large inclusive "churches," likely were preserved only in true churches.

But true churches multiplied greatly with revival of God's work at the Received-Text advent. In God's plan, true readings dropped in most Traditional-Text manuscripts would be renewed in the Received Text to extend the inerrant Traditional Text of earlier true churches to later true ones. This subject is discussed in detail in another IFBC booklet [14] and on www.go2ifbc.com

Conclusion: The items above are a tiny sampling of providential consistent KJB accuracy that can be presented,[14] and readers are wise to study the issues to ensure they have God's inerrant Word, not men's productions, in order to hear one day, *Well done, thou good and faithful servant...enter thou into the joy of thy Lord.*

[14]. L. Bednar. *The Case for Preservation of God's Inerrant Word.*
IFBC/WCBS booklet: Contact: 574-831-3770

Appendix

Dating of Job: Job was written ~2000 B.C, preceding the Pentateuch, for it doesn't mention Israel, and it shows the man Job personally administering the animal sacrifice of Abel and Noah, preceding a priestly-administered one of Moses. It says nothing of Abraham, greatest figure of his time, and Job was, *greatest of all the men of the east* (1:3), so the book was likely written prior to Abraham's time, ~2000 B.C. The name Uz, Job's land, is very old, deriving from Noah's great-grandson (Gen.10:21-23). Job preceded Abraham in that Job's life span was the 140 years he lived after his trials (42:16), plus preceding years in which he grew to manhood and begat 10 children old enough to sin (1:1-5), for a likely total of ~200 years. This is comparable to the 205 years of Abraham's father, Terah, and greater than the 175 years of Abraham, the 147 of Jacob and the 110 of Jacob's son Joseph. This places Job close to Terah's time in history as the life-span steadily decreased from a near-1000 year pre-flood level.

Yet another indicator that Job is the first book is its description of ancient dinosaurs, a matter discussed elsewhere (ref.14). And logic indicates Job was first, for it deals with the foremost human dilemma, the question of why hurt and evil dominate life, and it presents the solution, the Savior (Job 19:25-27).

Now we must account for the fact that names of Chaldean and Sabean tribes and the tribe of the friend Elihu link to descendants of Abraham and Nahor. Actually the name *Chaldean* preceded Abraham's time, for his first homeland was Ur of the Chaldees (Gen.11:31), and the name *Sabean* originated very early with Seba, Noah's great grandson (Gen.10:1-7). And the Naamathite tribal name of Job's friend Zophar, originated very early with the name of Cain's great granddaughter (Gen.4:22).

We must also account for the fact that Job's friends Eliphaz and Bildad have personal and/or tribal names identified with descendants of Abraham (Esau, Shuah, Jokshan - Gen.25:2) and his brother Nahor (Gen.22:20,21). Actually an inexact genealogy correlation indicates that names in lines of descent of Abraham and Nahor are repeat uses of names in Job. For example Eliphaz in Job is called a Temanite, while 1 Chronicles 1:36 says Teman in Abraham's line was a son of Eliphaz; a father can't continue a clan of which his son is patriarch, so Eliphaz in Abraham's line is a different Eliphaz. And Elihu the Buzite of Job has a father Barachel not noted in the line of Nahor, father of Buz and Huz (Gen.22:21), and Job's Ram in the line of Buz isn't in Nahor's line, the one other Ram in scripture being in Abraham's line (1 Chron.2:9 - Aram in Nahor's line and Ram are different Hebrew names). And Nahor's Huz and Buz are names likely derived from Job's land Uz, reflecting a trend in Nahor's clan in that his name derived from the name of his town and his brother Haran's name derived from the district of the town of Nahor. Job could be a son of Jacob's son Issachar (Gen.46:13), placing him around 1800-1700 B.C, but this is likely another use of the name Job, for nothing ties this man to Job's greatness in the Job book.

Repeat use of names in descent lines of Job and friends, in a somewhat-similar order, is no surprise. Children have long been named after Bible figures. As Abraham was the greatest man in his age, Job was in his, and children would be named after names in Job, this being the only Bible book at Abraham's time and for the next 500 years. Similar order likely reflects honoring of the first Bible book, and descendants of Abraham and Nahor would likely want their children's names to reflect the Job book due to great faith in God common to Job and Abraham. Inaccuracy in repeating names in a line would be due to conflicting human preferences, very limited Bible-manuscript availability so early in history and human uncertainty in the absence of early initial record-keeping. Record-keeping evidently began to be established with the book of Numbers and continued with Chronicles, Ezra and Nehemiah

Made in the USA
Charleston, SC
17 July 2012